Relativistic Causality
Lorentz transformation in Special Relativity

Aleks Kleyn

E-mail address: Aleks_Kleyn@MailAPS.org
URL: http://sites.google.com/site/alekskleyn/
URL: http://arxiv.org/a/kleyn_a_1
URL: http://AleksKleyn.blogspot.com/

ABSTRACT. The assertion about the possibility of motion faster than light does not contradict the special relativity. In order to develop the special relativity, is sufficient to assume independence of the speed of light on the reference frame. From equations of special relativity, it follows that object moving faster than light in vacuum cannot be carrier of relativistic causal relationship.

In the reference frame S_3, moving with superluminal speed $v_{3\cdot 2}$ relative to the reference frame S_2, temporal and spatial axes are swapped. Therefore, causal relationships in reference frames S_2 and S_3 are different. If a reference frame S_1 moves with speed $v_{1\cdot 2}$

$$\frac{c^2}{v_{3\cdot 2}} < v_{1\cdot 2} < c$$

relative to the reference frame S_2 in the direction of the increasing x, then the reference frame S_3 moves in the direction of decreasing values of x of the reference frame S_1. Therefore, observers of reference frame S_1 and S_2 observe differently the movement of the reference frame S_3.

In the paper, I considered the procedure of tracking of movement in the special relativity, as well I analyzed differences in motion with constant velocity, which is less than speed of light in vacuum, and motion with constant velocity, which is greater than the speed of light in vacuum.

Contents

1. Question of the Possibility of Superluminal Speed

Does the possibility of a physical object to move with a velocity greater than speed of light contradict the basic postulates of special and general relativity. No, it is not. The only postulate, which is in the basis of special relativity is the statement on the independence of the speed of light on the choice of reference frame. Lorentz transformations represented in the form

$$t_2 = \sqrt{\frac{c^2}{c^2 - v^2}} \left(t_1 - \frac{v}{c^2} x_1 \right)$$

$$x_2 = \sqrt{\frac{c^2}{c^2 - v^2}} (x_1 - vt_1)$$

are meaningful only if the relative speed is less than the speed of light. From this we can conclude that relative speed cannot be more than speed of light.

Recent experiment performed in CERN to measure neutrino velocity ([9]) resumed question about possibility of movement with speed larger than speed of light; as well the question arises: how it may affect the causal relationship? In this paper, I tried to answer these questions in the frame of special relativity.[1]

To understand law of transformation of coordinates during the transition to a reference frame which moves with the speed exceeding the speed of light, I turned directly to Einstein's papers [1, 2]. Before considering the transformation of coordinates during the transition to a reference frame that moves with the speed exceeding the speed of light, I considered the transformation of coordinates during the transition to a reference frame that moves with the speed less than the speed of light.

From the laws of the dynamics of the special relativity, it follows that if there are no external actions on physical object, then either in any reference frame the object moves with speed of light, or there is no reference frame relative to which the object in question moves with speed of light.

Analysis of the transformation shows that in the reference frame S_3, moving with superluminal speed relative to the reference frame S_1, temporal and spatial axes are swapped. This means that, in the reference frame S_3, familiar causal relationship in time is destroyed, however causal relationship along the spatial axis appears.

If the reference frame S_3 moves with velocity $v_{3.2}$ in the direction of increasing values of x of the reference frame S_2 and the reference frame S_2 moves relative to the reference frame S_1 with speed $v_{2.1}$

$$-c < v_{2.1} < c$$

then speed of the reference frame S_3 relative to the reference frame S_2 also is greater than the speed of light. However, according to the corollary 5.6, if the reference frame S_1 moves with speed $v_{1.2}$

$$\frac{c^2}{v_{3.2}} < v_{1.2} < c$$

relative to the reference frame S_2 in the direction of the increasing x, then the reference frame S_3 moves in the direction of decreasing values of x of the reference frame S_1. Therefore, observers of reference frame S_1 and S_2 observe differently the consequence of spatial events in the reference frame S_3. If the object moves relative to the reference frame S faster than light in vacuum, then this object cannot be carrier of causal relationship of the reference frame S.

---

[1]During XX century, numerous researches confirmed special relativity. The goal of new theories that appeared recently is to join concepts of quantum mechanics and general relativity. These theory use geometry which is different from geometry of event space of special relativity. Correspondingly, there exist a few different concepts of causal relationship. Each of these concepts has its own area of application. However, either we expect that the predictions of the new theory under certain conditions are close to predictions of special or general relativity, or we can see difference in experiment.

2. Theorems about Reflection

The event space of special relativity is not Euclidean space. So we will not use the theorem considered in this section to prove any statement about event space. However we need these theorems to understand why we selected a particular model.

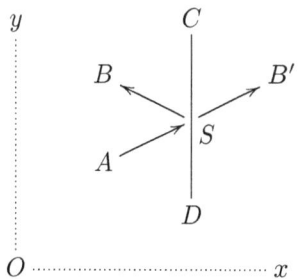

THEOREM 2.1. *Let the straight line CD in the plane xy be parallel to axis of y. Let the ball move in the plane xy along the ray AS and is reflected by the line CD at the point S. Let the reflected ball move along the ray SB. Since the ray AS is not parallel to axis of x and is not parallel to axis of y, then coordinate y changes monotonically and coordinate x has an extremum at the point S.*

REMARK 2.2. Consider cases that we did not listed in the theorem.

If the point A belongs to the line CD, then the point B also belongs to the line CD. Then coordinate y changes monotonically and coordinate x is constant; so any point can be considered as extremum of x.

Since the ray AS is parallel to axis of x, then the ray SB is also parallel to axis of x. Then coordinate x has an extremum at the point S and coordinate y is constant; so we can assume that y changes monotonically.

REMARK 2.3. The angle ASD is called angle of incidence. The angle CSB is called angle of reflection. Since the ray AS is not parallel to axis of x, then the angle of incidence is not right angle. Since the ray AS is not parallel to axis of y, then the value the angle of incidence does not equal 0. If the angle ASD is obtuse angle, then we can consider the adjacent angle as the angle of incidence. So in the following we assume that angle of incidence is acute.

PROOF. The ray SB' extends the ray AS. If there is no reflection at the point S, the ball moves from the point A into the point B'. Under this assumption, coordinates x, y change monotonically. As a result of reflection, the ball moves along the ray SB symmetric the ray SB' with respect to the line CD. So the speed of change of coordinate y does not change, and the speed of change of coordinate x changes sign at the point S. Therefore, coordinate x has an extremum at the point S. $\square$

THEOREM 2.4. *Let the straight line CD in the plane xy be parallel to axis of y. Let the ball move in the plane xy along the ray AS and is reflected by the straight line CD at the point S. Let the reflected ball move along the ray SB. Let the ray AS be not parallel to axis of x and is not parallel to axis of y. Let the value of angle between the straight line KL and the straight line CD is less than the value of the angle ASD. Since the straight line LK intersects the ray AS at the point E, then the straight line LK intersects the ray SB at the point F. Under this assumption, the coordinate y_S is between coordinates y_E and y_F.*

PROOF. According to the remark 2.3, the angle ASD is acute angle.

Let the straight line KL is parallel to the straight line CD. Since $SB = AS$, then
$$\angle BAS = \angle ABS$$
Since
$$\angle ASD = \angle BSC$$
then
$$\angle ASD = \angle BAS$$
Therefore
$$AB \parallel CD \quad AB \parallel KL$$

Since the straight line KL intersects segment AS at the point E, then, according to Pasch's axiom the straight line KL must intersect either segment AB, or segment BS. Since $AB \parallel KL$, then the straight line KL intersects segment BS at the point F.

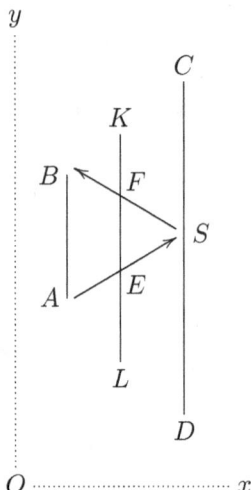

Let $\angle LKD$ be acute angle. Let $EG \perp CD$. Since $\angle ASD$, $\angle LKD$ are acute angles, then points S, K and the point D are on opposite sides of the point G. Since
$$\angle LKD < \angle ASD$$
then the point K is farther from the point G than the point S. Since angles LKD and BSC are acute angles, then there exists triangle KSF, whose vertex F is the intersection of the line KL and the ray SB.

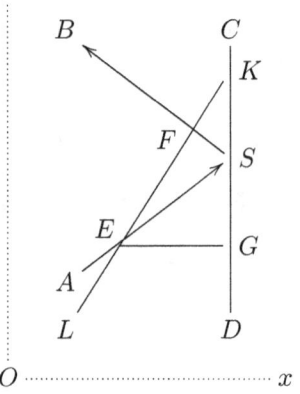

Let $\angle LKC$ be acute angle. Let $EG \perp CD$. Since $\angle ASD$, $\angle LKC$ are acute angles, then points S and K are on opposite sides of the point G. Since sum of angles LKC and BSD is less than $180°$, then there exists triangle KSF, whose vertex F is the intersection of the line KL and the ray SB.

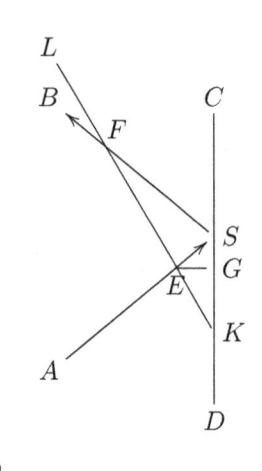

The statement that the coordinate y_S is between coordinates y_E and y_F follows from the theorem 2.1. $\qquad\square$

3. Lorentz Transformation, Relative Speed is Less than Speed of Light

Consider event space of special relativity.[2] For the spatial measurement we need three rigid rods which are mutually perpendicular and rigidly connected, as well we need rigid unit of measure. Given three rods and unit of measure generate Cartesian coordinate system where coordinates x, y, z uniquely determine the position of a point in space. To measure time of point event, we need clock which is at rest relative to coordinate system in immediate vicinity of the point event.

Let clocks at rest relative to coordinate system be located at different points. Let these clocks be equivalent, i. e., the difference of readings of any two such clocks remains unchanged. In order for these clocks give the time in the form in which it is necessary for the purposes of physical objectives, clock should be verified such way that the speed of light propagation in empty space is equal to a universal constant c provided that the coordinate system is a non-accelerated. The set of readings of calibrated such way clocks is called the time of coordinate system.

The coordinate system together with unit of measure and clocks which serve to establish the time of the system is called reference frame S. We will identify any event in reference frame S using coordinates x, y, z and time t. The set of events in reference frame S forms event space. Hereinafter we will identify the observer and origin of reference frame accompanying him. The trajectory of observer in event space is called world line of observer or world line of reference frame.

The straight line in the event space corresponds to motion with constant velocity. The tangent of the angle between this line and the axis of t is equal to the speed. Straight lines l_1, l_2 describe the trajectory of light passing through the origin. The straight line b describes the trajectory of motion with speed less than the speed of light. The angle between the straight line b and the axis of t is less than the angle between the straight line l_1 and the axis of t.

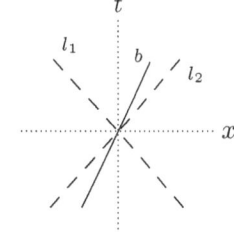

Let the observer A_2 move along the axis of x with speed $v < c$ relative to the observer A_1 (the straight line AB in the event space). At the time t_1 (the point F_1 in the event space), the observer A_2 emits a ray in direction of the point with coordinate x_2 (the point F_2 in the event space). At the point with coordinate x_2 there is mirror which reflect the ray at the time t_2. In the event space, reflection of light at the point with coordinate x_2 is equivalent to reflection by the straight line CD which is parallel to the axis of t. Therefore, we can use theorems 2.1, 2.4. According to the theorem 2.4, the reflected ray intersects the straight line AB at the point F_3 at time t_3, $t_1 < t_2 < t_3$.

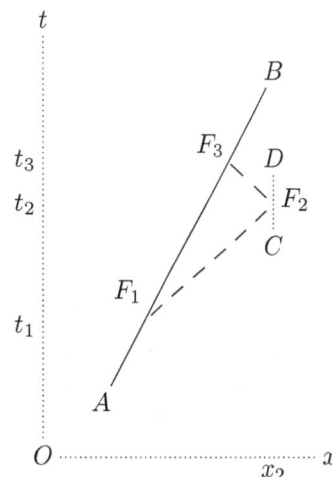

Consider two reference frames in the space.[3] Let the axis of x of the two reference frames coincide. Let the axes of y and z of the two reference frames be parallel. Let the length and time scales in the two reference frames coincide.

---

[2]The definition of reference frame in this section follows the definition of reference frame in [2], §1.
[3]In this section, I follow [1], §3, [2], §3.

We say that the reference frame S_2 has a constant velocity relative to reference frame S_1, if the origin of the reference frame S_2 has a constant velocity relative to the origin of the reference frame S_1 and this velocity is transmitted to the axes of the coordinates and the relevant scales.

Let the reference frame S_2 have a constant velocity $v < c$ in the direction of the increasing x of the reference frame S_1. To any system of values t_1, x_1, y_1, z_1, which completely defines the place and time of an event in the reference frame S_1, there belongs a system of values t_2, x_2, y_2, z_2, determining that event relatively to the reference frame S_2. The equations connecting these quantities must be linear on account of the homogeneity of space and time.

If $v = 0$, then world lines of reference frames S_1 and S_2 are parallel. The change of coordinates in the transition from the reference frame S_1 to the reference frame S_2 is the parallel shift generated by the change of space and time origins. If $v \neq 0$, then world lines of reference frames S_1 and S_2 have a unique point of intersection. Let choose as time origin in reference frames S_1 and S_2 the moment of intersection of world lines of reference frames S_1 and S_2. Then equations of the linear transformation are homogeneous.

If we place

$$(3.1) \qquad x'_1 = x_1 - vt_1$$

then a point at rest in the reference frame S_2 must have a system of values x'_1, y_1, z_1, independent of time. Therefore, t_2 is linear function of x'_1, y_1, z_1, t_1

$$(3.2) \qquad t_2 = Ax'_1 + Bt_1$$

It is assumed that coefficients A, B may depend on the speed v. It is assumed that

$$(3.3) \qquad y_2 = y_1 \quad z_2 = z_1$$

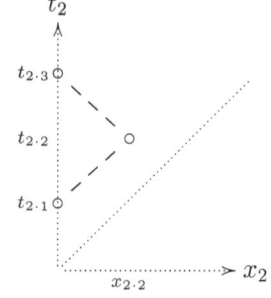

Let a ray be emitted at the time $t_{2 \cdot 1}$ along the x-axis from the origin of the reference frame S_2 to $x_2 = x_{2 \cdot 2}$, and at the time $t_2 = t_{2 \cdot 2}$ this ray be reflected thence to the origin of the reference frame, arriving there at the time $t_{2 \cdot 3}$. Then

$$(3.4) \qquad t_{2 \cdot 2} = \frac{t_{2 \cdot 1} + t_{2 \cdot 3}}{2}$$

If the point is on the world line of the reference frame S_2, then $x'_1 = 0$. In particular,

$$(3.5) \qquad x'_{1 \cdot 1} = x_{1 \cdot 1} - vt_{1 \cdot 1} = 0$$

$$(3.6) \qquad x'_{1 \cdot 3} = x_{1 \cdot 3} - vt_{1 \cdot 3} = 0$$

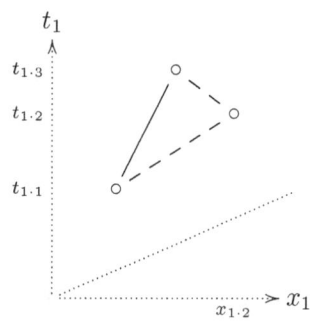

According to the definition (3.1), the value $x'_1 = x'_{1 \cdot 2}$ in the reference frame S_1 corresponds to the point $x_2 = x_{2 \cdot 2}$ at rest in the reference frame S_2, i.e. this value corresponds to the point which have a constant velocity v in the direction of the increasing x and which at the time $t_1 = 0$ is located in a point $x_1 = x'_{1 \cdot 2}$. Therefore, a ray emitted at the point $x_1 = x_{1 \cdot 1}$ at time $t_1 = t_{1 \cdot 1}$, arrives at the point corresponding to the value $x'_1 = x'_{1 \cdot 2}$ at time $t_1 = t_{1 \cdot 2}$ such that

$$(3.7) \qquad x_{1 \cdot 1} + c(t_{1 \cdot 2} - t_{1 \cdot 1}) = x'_{1 \cdot 2} + vt_{1 \cdot 2}$$

From the equation (3.7), it follows that

(3.8) $$x_{1.1} - vt_{1.1} + c(t_{1.2} - t_{1.1}) = x'_{1.2} + vt_{1.2} - vt_{1.1}$$

From equations (3.5), (3.8), it follows that

(3.9) $$c(t_{1.2} - t_{1.1}) = x'_{1.2} + v(t_{1.2} - t_{1.1})$$

From the equation (3.9), it follows that

(3.10) $$t_{1.2} = t_{1.1} + \frac{x'_{1.2}}{c - v}$$

From the equation (3.6), it follows that a ray reflected at the point $x_1 = x_{1.2}$ at time $t_1 = t_{1.2}$, arrives at the point corresponding to the value $x'_1 = x'_{1.3}$ at time $t_1 = t_{1.3}$ such that

(3.11) $$(x'_{1.2} + vt_{1.2}) - c(t_{1.3} - t_{1.2}) = vt_{1.3}$$

From the equation (3.11), it follows that

(3.12) $$x'_{1.2} - c(t_{1.3} - t_{1.2}) = vt_{1.3} - vt_{1.2} = v(t_{1.3} - t_{1.2})$$

From the equation (3.12), it follows that

(3.13) $$x'_{1.2} = (c + v)(t_{1.3} - t_{1.2})$$

From the equation (3.13), it follows that

(3.14) $$t_{1.3} = t_{1.2} + \frac{x'_{1.2}}{c + v}$$

From equations (3.10), (3.14), it follows that

(3.15) $$t_{1.3} = t_{1.1} + \frac{x'_{1.2}}{c - v} + \frac{x'_{1.2}}{c + v}$$

From equations (3.2), (3.4), (3.10), (3.15), it follows that

(3.16) $$2\left(Ax'_{1.2} + B\left(t_{1.1} + \frac{x'_{1.2}}{c - v}\right)\right) = Bt_{1.1} + B\left(t_{1.1} + \frac{x'_{1.2}}{c - v} + \frac{x'_{1.2}}{c + v}\right)$$

From the equation (3.16), it follows that

(3.17) $$A = \frac{B}{2}\left(\frac{1}{c + v} - \frac{1}{c - v}\right) = -B\frac{v}{c^2 - v^2}$$

From equations (3.2), (3.17), it follows that

(3.18) $$t_{2.2} = B\left(t_{1.2} - \frac{v}{c^2 - v^2}x'_{1.2}\right)$$

According to the construction, $x'_{1.2}$ in the equation (3.18) is arbitrary. Similarly, $t_{1.2}$ is arbitrary, since according to the equation (3.10), for the given value $x'_{1.2}$ we can find such value $t_{1.1}$ that we get given value $t_{1.2}$. So, we can write the equation (3.18) as follows

(3.19) $$t_2 = B\left(t_1 - \frac{v}{c^2 - v^2}x'_1\right)$$

In particular, from equations (3.10), (3.19), it follows that

(3.20) $$t_{2.1} = Bt_{1.1}$$

Since light is also propagated with velocity c when measured in the reference frame S_2, then from the equations (3.19), (3.20), it follows that[4]

(3.21)
$$x_2 = c(t_2 - t_{2\cdot1}) = cB\left(t_1 - \frac{v}{c^2 - v^2}x_1' - t_{1\cdot1}\right)$$

From equations (3.21), (3.10), it follows that

(3.22)
$$x_2 = cB\left(t_1 - t_{1\cdot1} - \frac{v}{c^2 - v^2}x_1'\right) = cB\left(\frac{x_1'}{c-v} - \frac{v}{c^2-v^2}x_1'\right)$$

From equations (3.1), (3.18), (3.22) it follows that

(3.23)
$$t_2 = B\left(t_1 - \frac{v}{c^2-v^2}(x_1 - vt_1)\right) = B\frac{c^2}{c^2-v^2}\left(t_1 - \frac{v}{c^2}x_1\right)$$
$$x_2 = B\frac{c^2}{c^2-v^2}x_1' \qquad\qquad = B\frac{c^2}{c^2-v^2}(x_1 - vt_1)$$

Since the light propagation velocity in empty space[5] with respect to reference frames S_1 and S_2 equals c, following equations

(3.24)
$$x_1^2 + y_1^2 + z_1^2 = c^2 t_1^2$$
$$x_2^2 + y_2^2 + z_2^2 = c^2 t_2^2$$

must be equivalent. From equations (3.3), (3.24), it follows that

(3.25)
$$x_1^2 - c^2 t_1^2 = x_2^2 - c^2 t_2^2$$

From equations (3.23), (3.25), it follows that

(3.26)
$$x_1^2 - c^2 t_1^2$$
$$= B^2\frac{c^4}{(c^2-v^2)^2}(x_1 - vt_1)^2 - c^2 B^2\frac{c^4}{(c^2-v^2)^2}\left(t_1 - \frac{v}{c^2}x_1\right)^2$$
$$= \frac{B^2}{c^2}\frac{c^4}{(c^2-v^2)^2}(c^2(x_1 - vt_1)^2 - (c^2 t_1 - vx_1)^2)$$
$$= B^2\frac{c^2}{(c^2-v^2)^2}(c^2 x_1^2 - 2c^2 vx_1 t_1 + c^2 v^2 t_1^2 - c^4 t_1^2 + 2c^2 vx_1 t_1 - v^2 x_1^2)$$
$$= B^2\frac{c^2}{(c^2-v^2)^2}((c^2 - v^2)x_1^2 - c^2(c^2 - v^2)t_1^2)$$
$$= B^2\frac{c^2}{c^2-v^2}(x_1^2 - c^2 t_1^2)$$

From the equation (3.26), it follows that

(3.27)
$$B = \pm\sqrt{\frac{c^2 - v^2}{c^2}}$$

To determine sign in the equation (3.27) we set $v = 0$. Then reference frames S_1 and S_2 coincide. From the equation (3.23) it follows that

(3.28)
$$t_2 = Bt_1$$
$$x_2 = Bx_1$$

---

[4]Technically, we have to write
$$x_2 = x_{2\cdot1} + c(t_2 - t_{2\cdot1})$$
However, according to construction $x_{2\cdot1} = 0$.

[5]In this part of the reasoning, I follow [2], §3.

$B = 1$ follows from the equation (3.28). From the equation (3.27), it follows that

(3.29)
$$B = \sqrt{\frac{c^2 - v^2}{c^2}}$$

From the equations (3.23), (3.29), it follows that

(3.30)
$$t_2 = \sqrt{\frac{c^2}{c^2 - v^2}}\left(t_1 - \frac{v}{c^2}x_1\right)$$
$$x_2 = \sqrt{\frac{c^2}{c^2 - v^2}}(x_1 - vt_1)$$

From the equation (3.30) it follows that

(3.31)
$$t_1 = \sqrt{\frac{c^2}{c^2 - v^2}}\left(t_2 + \frac{v}{c^2}x_2\right)$$
$$x_1 = \sqrt{\frac{c^2}{c^2 - v^2}}(x_2 + vt_2)$$

4. Lorentz Transformation, Relative Speed is Greater than Speed of Light

Consider two reference frames in the space.[6] Let the axes of x of the two reference frames coincide. Let the axes of y and z of the two reference frames be parallel. Let the length and time scales in the two reference frames coincide.

To any system of values t_1, x_1, y_1, z_1, which completely defines the place and time of an event in the reference frame S_1, there belongs a system of values t_2, x_2, y_2, z_2, determining that event relatively to the reference frame S_2. The equations connecting these quantities must be linear on account of the homogeneity of space and time.

Let reference frame S_2 have a constant velocity $v > c$ in the direction of the increasing x of the reference frame S_1 (the straight line AB in the event space). At the time $t_{1.1}$ (the point F_1 in the event space), the observer A_2 emits a ray in direction of the point with coordinate $x_1 = x_{1.2}$ (the point F_2 in the event space). At the point with coordinate $x_1 = x_{1.2}$ there is mirror which reflect the ray at the time $t_1 = t_{1.2}$.

In the event space, reflection of light at the point with coordinate x_2 is equivalent to reflection by the straight line CD which is parallel to the axis of t. However we cannot directly apply the method used in section 3, since the light signal reflected at the point $x_{1.1}$, never meet the observer associated with the reference frame S_2.

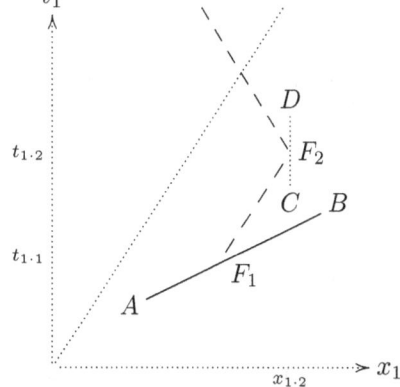

---

[6]In this section, I follow [1], §3.

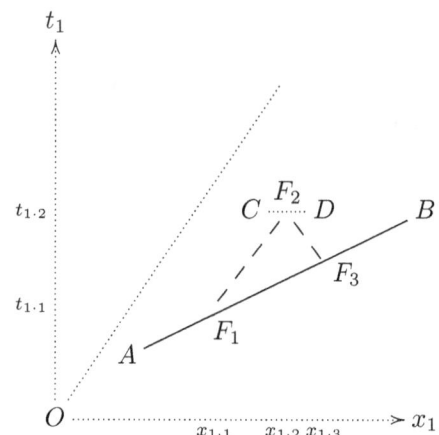

According to [5], p. 94, 98, photon with positive frequency propagates forward in time and photon with negative frequency propagates backward in time. We assume that photon with negative frequency is emitted at the point $x_{1\cdot 2}$ and propagates along the axis of x in the same direction as original photon (the straight line $F_2 F_3$ in the event space). This is equivalent to the statement that in the event F_2 photon is reflected by the straight line CD which is parallel to the axis of x.

Since the angle between the straight line AB and the axis of t is larger than the angle between the straight line $F_1 F_2$ and the axis of t, then the angle between the straight line AB and the axis of x is less than the angle between the straight line $F_1 F_2$ and the axis of x. According to the theorem 2.4, photon with negative frequency emitted at point F_2 intersects the straight line AB at point F_3

$$x_{1\cdot 1} < x_{1\cdot 2} < x_{1\cdot 3}$$

According to the theorem 2.1,

$$t_{1\cdot 1} < t_{1\cdot 2} \quad t_{1\cdot 3} < t_{1\cdot 2}$$

Therefore, for reference frame S_2 the coordinate x plays role of time and the coordinate t is the spatial coordinate.

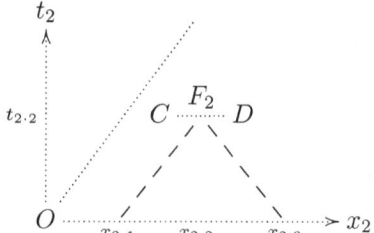

Since the coordinate t in reference frame S_2 may change arbitrary, then the observer in reference frame S_2 assumes that at rest $t = 0$. Therefore, reflected ray arrives at the origin of the reference frame S_2.

If we place

(4.1)
$$t_1' = t_1 - \frac{x_1}{v}$$

then a point at rest in the reference frame S_2 must have a system of values t_1', y_1, z_1, independent of x_1. Therefore, x_2 is linear function of t_1', y_1, z_1, x_1

(4.2)
$$x_2 = A x_1 + B t_1'$$

It is assumed that coefficients A, B may depend on the speed v. It is assumed that

(4.3)
$$y_2 = y_1 \quad z_2 = z_1$$

Let a ray be emitted at the time $x_{2\cdot 1}$ along the t-axis from the origin of the reference frame S_2 to $t_{1\cdot 2}$, and at the time $x_{2\cdot 2}$ be reflected thence to the origin of the reference frame, arriving there at the time $x_{2\cdot 3}$. Then

(4.4)
$$x_{2\cdot 2} = \frac{x_{2\cdot 1} + x_{2\cdot 3}}{2}$$

If the point is on the world line of the reference frame S_2, then $t'_1 = 0$. In particular,

(4.5)
$$t'_{1\cdot1} = t_{1\cdot1} - \frac{x_{1\cdot1}}{v} = 0$$

(4.6)
$$t'_{1\cdot3} = t_{1\cdot3} - \frac{x_{1\cdot3}}{v} = 0$$

According to the definition (4.1), the value $t'_1 = t'_{1\cdot2}$ in the reference frame S_1 corresponds to the point $t_2 = t_{2\cdot2}$ at rest in the reference frame S_2, i.e. this value corresponds to the point which have a constant velocity v in the direction of the increasing x and which at the time $x_1 = 0$ is located in a point $t_1 = t'_{1\cdot2}$. Therefore, a ray emitted at the point $x_1 = x_{1\cdot1}$ at time $t_1 = t_{1\cdot1}$, arrives at the point corresponding to the value $t'_1 = t'_{1\cdot2}$ at time $x_1 = x_{1\cdot2}$ such that

(4.7)
$$t_{1\cdot1} + \frac{x_{1\cdot2} - x_{1\cdot1}}{c} = t'_{1\cdot2} + \frac{x_{1\cdot2}}{v}$$

From the equation (4.7), it follows that

(4.8)
$$t_{1\cdot1} - \frac{x_{1\cdot1}}{v} + \frac{x_{1\cdot2} - x_{1\cdot1}}{c} = t'_{1\cdot2} + \frac{x_{1\cdot2}}{v} - \frac{x_{1\cdot1}}{v}$$

From equations (4.5), (4.8), it follows that

(4.9)
$$\frac{x_{1\cdot2} - x_{1\cdot1}}{c} = t'_{1\cdot2} + \frac{x_{1\cdot2} - x_{1\cdot1}}{v}$$

From the equation (4.9), it follows that

(4.10)
$$x_{1\cdot2} = x_{1\cdot1} + t'_{1\cdot2}\frac{cv}{v - c}$$

From the equation (4.6), it follows that a ray reflected at the point $x_1 = x_{1\cdot2}$ at time $t_1 = t_{1\cdot2}$, arrives at the point corresponding to the value $t'_1 = t'_{1\cdot3}$ at time $x_1 = x_{1\cdot3}$ such that

(4.11)
$$\left(t'_{1\cdot2} + \frac{x_{1\cdot2}}{v}\right) - \frac{x_{1\cdot3} - x_{1\cdot2}}{c} = \frac{x_{1\cdot3}}{v}$$

From the equation (4.11), it follows that

(4.12)
$$t'_{1\cdot2} - \frac{x_{1\cdot3} - x_{1\cdot2}}{c} = \frac{x_{1\cdot3}}{v} - \frac{x_{1\cdot2}}{v} = \frac{x_{1\cdot3} - x_{1\cdot2}}{v}$$

From the equation (4.12), it follows that

(4.13)
$$t'_{1\cdot2} = \frac{c + v}{cv}(x_{1\cdot3} - x_{1\cdot2})$$

From the equation (4.13), it follows that

(4.14)
$$x_{1\cdot3} = x_{1\cdot2} + t'_{1\cdot2}\frac{cv}{c + v}$$

From equations (4.10), (4.14), it follows that

(4.15)
$$x_{1\cdot3} = x_{1\cdot1} + t'_{1\cdot2}\frac{cv}{v - c} + t'_{1\cdot2}\frac{cv}{c + v}$$

From equations (4.2), (4.4), (4.10), (4.15), it follows that

(4.16)
$$2\left(Bt'_{1\cdot2} + A\left(x_{1\cdot1} + t'_{1\cdot2}\frac{cv}{v - c}\right)\right)$$
$$= Ax_{1\cdot1} + A\left(x_{1\cdot1} + t'_{1\cdot2}\frac{cv}{v - c} + t'_{1\cdot2}\frac{cv}{c + v}\right)$$

From the equation (4.16), it follows that

(4.17)
$$B = \frac{A}{2}\left(\frac{cv}{c + v} - \frac{cv}{v - c}\right) = -A\frac{c^2v}{v^2 - c^2}$$

From equations (4.2), (4.17), it follows that

(4.18)
$$x_{2\cdot 2} = A\left(x_{1\cdot 2} - \frac{c^2 v}{v^2 - c^2}t'_{1\cdot 2}\right)$$

According to the construction, $t'_{1\cdot 2}$ in the equation (4.18) is arbitrary. Similarly, $x_{1\cdot 2}$ is arbitrary, since according to the equation (4.10), for the given value $t'_{1\cdot 2}$ we can find such value $x_{1\cdot 1}$ that we get given value $x_{1\cdot 2}$. So, we can write the equation (4.18) as follows

(4.19)
$$x_2 = A\left(x_1 - \frac{c^2 v}{v^2 - c^2}t'_1\right)$$

In particular, from equations (4.10), (4.19), it follows that

(4.20)
$$x_{2\cdot 1} = Ax_{1\cdot 1}$$

Since light is also propagated with velocity c when measured in the reference frame S_2, then from the equations (4.19), (4.20), it follows that[7]

(4.21)
$$t_2 = (x_2 - x_{2\cdot 1})c = Ac\left(x_1 - \frac{c^2 v}{v^2 - c^2}t'_1 - x_{1\cdot 1}\right)$$

From equations (4.21), (4.10), it follows that

(4.22)
$$t_2 = Ac\left(x_1 - x_{1\cdot 1} - \frac{c^2 v}{v^2 - c^2}t'_1\right) = Ac\left(\frac{cv}{v - c}t'_1 - \frac{c^2 v}{v^2 - c^2}t'_1\right)$$

From equations (4.1), (4.18), (4.22), it follows that

(4.23)
$$x_2 = A\left(x_1 - \frac{c^2 v}{v^2 - c^2}\left(t_1 - \frac{x_1}{v}\right)\right) = A\frac{v}{v^2 - c^2}(vx_1 - c^2 t_1)$$
$$t_2 = Ac\frac{cv^2}{v^2 - c^2}\left(t_1 - \frac{x_1}{v}\right) \qquad = A\frac{c^2 v}{v^2 - c^2}(vt_1 - x_1)$$

Since the light propagation velocity in empty space[8] with respect to reference frames S_1 and S_2 equals c, following equations

(4.24)
$$x_1^2 + y_1^2 + z_1^2 = c^2 t_1^2$$
$$t_2^2 + y_2^2 + z_2^2 = c^2 x_2^2$$

must be equivalent. From equations (4.3), (4.24), it follows that

(4.25)
$$x_1^2 - c^2 t_1^2 = t_2^2 - c^2 x_2^2$$

---

[7]The coordinate x_2 is used to measure time in reference frame S_2; and the coordinate t_2 is used to measure spatial intervals in reference frame S_2. Maxwell equations are valid also in reference frame S_2. In particular, the speed of light in vacuum, measured in reference frame S_2, is also equal to c. Technically, we have to write
$$t_2 = t_{2\cdot 1} + (x_2 - x_{2\cdot 1})c$$
However, according to construction $t_{2\cdot 1} = 0$.

[8]In this part of the reasoning, I follow [2], §3.

From equations (4.23), (4.25), it follows that

$$x_1^2 - c^2 t_1^2$$

$$= A^2 \frac{c^4 v^2}{(v^2 - c^2)^2}(vt_1 - x_1)^2 - c^2 A^2 \frac{v^2}{(v^2 - c^2)^2}(vx_1 - c^2 t_1)^2$$

$$= A^2 \frac{c^2 v^2}{(v^2 - c^2)^2}(c^2(vt_1 - x_1)^2 - (vx_1 - c^2 t_1)^2)$$

(4.26)
$$= A^2 \frac{c^2 v^2}{(v^2 - c^2)^2}(c^2 v^2 t_1^2 - 2c^2 vt_1 x_1 + c^2 x_1^2 - v^2 x_1^2 + 2vc^2 x_1 t_1 - c^4 t_1^2)$$

$$= A^2 \frac{c^2 v^2}{(v^2 - c^2)^2}((c^2 - v^2)x_1^2 - c^2(c^2 - v^2)t_1^2)$$

$$= A^2 \frac{c^2 v^2}{v^2 - c^2}(x_1^2 - c^2 t_1^2)$$

From the equation (4.26), it follows that

(4.27)
$$A = \pm\sqrt{\frac{v^2 - c^2}{c^2 v^2}}$$

To determine sign in the equation (4.27) we put attention that in equation (4.23) coordinates x_1, x_2 increase simultaneously when $v > 0$. Therefore, we place

(4.28)
$$A = \sqrt{\frac{v^2 - c^2}{c^2 v^2}}$$

From the equations (4.23), (4.28), it follows that

(4.29)
$$t_2 = \sqrt{\frac{c^2}{v^2 - c^2}}(vt_1 - x_1)$$

$$x_2 = \sqrt{\frac{c^2}{v^2 - c^2}}\left(\frac{v}{c^2}x_1 - t_1\right)$$

From the equation (4.29) it follows that

(4.30)
$$t_1 = \sqrt{\frac{c^2}{v^2 - c^2}}\left(x_2 + \frac{v}{c^2}t_2\right)$$

$$x_1 = \sqrt{\frac{c^2}{v^2 - c^2}}(t_2 + vx_2)$$

5. Measurement of Speed

Let reference frame S_3 have a constant velocity $v_{3.2}$ in the direction of the increasing x of the reference frame S_2. Let choose as time origin in reference frame S_2 the moment of intersection of world lines of reference frames S_2 and S_3. Let choose as origin of coordinate x in reference frame S_2 the point of intersection of world lines of reference frames S_2 and S_3. Consider the procedure of measurement of speed of reference frame S_3 relative to reference frame S_2.[9]

Assume that the observer of reference frame S_2 placed at the point $x_2 = x_{2.1}$ a device to send the light rays to the observer. At the time when world line of reference frame S_3 crosses world line of reference frame S_2, the observer sends light ray to the point $x_2 = x_{2.1}$. The moment of arrival of the light ray to a point $x_2 = x_{2.1}$ is represented by the point F_1 on the diagram of event space. The installed device sends an observer light ray with frequency ω_1. The observer receives this light ray at time $t_{2.3}$. The moment of arrival

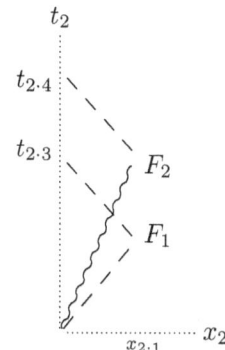

---

[9]We considered similar procedure in the section [10]-11.

of reference frame S_3 to a point $x_2 = x_{2 \cdot 1}$ is represented by the point F_2 on the diagram of event space. The installed device sends an observer light ray with frequency ω_2. The observer receives this light ray at time $t_{2 \cdot 4}$.

Since the observer of reference frame S_2 knows the time $t_{2 \cdot 3}$, then it follows from this that

(5.1)
$$x_{2 \cdot 2} = x_{2 \cdot 1} = \frac{t_{2 \cdot 3}}{2} c$$

Based on values $t_{2 \cdot 3}$ and $t_{2 \cdot 4}$, the observer of reference frame S_2 knows that the reference frame S_3 arrived to a point $x_2 = x_{2 \cdot 1}$ at the time

(5.2)
$$t_{2 \cdot 2} = t_{2 \cdot 4} - \frac{t_{2 \cdot 3}}{2}$$

From equations (5.1), (5.2), it follows that the speed of the reference frame S_3 relative to the reference frame S_2 is defined by value

(5.3)
$$v_{3 \cdot 2} = \frac{x_{2 \cdot 2}}{t_{2 \cdot 2}} = \frac{t_{2 \cdot 3} c}{2 t_{2 \cdot 4} - t_{2 \cdot 3}} = \frac{c}{2 \dfrac{t_{2 \cdot 4}}{t_{2 \cdot 3}} - 1}$$

From the equation (5.3), it follows that

$$t_{2 \cdot 3} < t_{2 \cdot 4} \Rightarrow v_{3 \cdot 2} < c$$

$$t_{2 \cdot 3} > t_{2 \cdot 4} \Rightarrow v_{3 \cdot 2} > c$$

Let the reference frame S_2 have a constant velocity $v_{2 \cdot 1}$

$$-c < v_{2 \cdot 1} < c$$

in the direction of the increasing x of the reference frame S_1. According to the equation (3.31), relative to the reference frame S_1, the point F_2 has coordinates

(5.4)
$$x_{1 \cdot 2} = \sqrt{\frac{c^2}{c^2 - v_{2 \cdot 1}^2}} (x_{2 \cdot 2} + v_{2 \cdot 1} t_{2 \cdot 2})$$

$$t_{1 \cdot 2} = \sqrt{\frac{c^2}{c^2 - v_{2 \cdot 1}^2}} \left(t_{2 \cdot 2} + \frac{v_{2 \cdot 1}}{c^2} x_{2 \cdot 2} \right)$$

From equations (5.3), (5.4), it follows that the speed of the reference frame S_3 relative to the reference frame S_1 is defined by value

(5.5)
$$v_{3 \cdot 1} = \frac{x_{1 \cdot 2}}{t_{1 \cdot 2}} = \frac{x_{2 \cdot 2} + v_{2 \cdot 1} t_{2 \cdot 2}}{t_{2 \cdot 2} + \dfrac{v_{2 \cdot 1}}{c^2} x_{2 \cdot 2}} = \frac{v_{3 \cdot 2} + v_{2 \cdot 1}}{1 + \dfrac{v_{3 \cdot 2} v_{2 \cdot 1}}{c^2}}$$

Signs of speed $v_{3 \cdot 1}$ and of speed $v_{3 \cdot 2}$ may coincide; but the case, when the signs are different, is more interesting for us. To understand the phenomena with which it is connected, we write equations (5.4) in the form

(5.6)
$$x_{1 \cdot 2} = \sqrt{\frac{c^2}{c^2 - v_{2 \cdot 1}^2}} \left(1 + v_{2 \cdot 1} \frac{t_{2 \cdot 2}}{x_{2 \cdot 2}} \right) x_{2 \cdot 2}$$

$$t_{1 \cdot 2} = \sqrt{\frac{c^2}{c^2 - v_{2 \cdot 1}^2}} \left(1 + \frac{v_{2 \cdot 1}}{c^2} \frac{x_{2 \cdot 2}}{t_{2 \cdot 2}} \right) t_{2 \cdot 2}$$

From equations (5.3), (5.6), it follows that

(5.7)
$$x_{1\cdot2} = \sqrt{\frac{c^2}{c^2 - v_{2\cdot1}^2}} \left(1 + \frac{v_{2\cdot1}}{v_{3\cdot2}}\right) x_{2\cdot2}$$

$$t_{1\cdot2} = \sqrt{\frac{c^2}{c^2 - v_{2\cdot1}^2}} \left(1 + \frac{v_{2\cdot1}v_{3\cdot2}}{c^2}\right) t_{2\cdot2}$$

THEOREM 5.1. *Let*

(5.8)
$$0 < v_{3\cdot2} < c$$

(5.9)
$$-c < v_{2\cdot1} < c$$

Since the speed $|v_{2\cdot1}|$ of the reference frame S_2 relative to the reference frame S_1 in the direction of the decreasing x is larger than the speed of the reference frame S_3 relative to the reference frame S_2, then the reference frame S_3 moves relative to the reference frame S_1 in the direction of the decreasing x.

PROOF. From the equation (5.7) and the inequalities (5.8), (5.9), it follows that $t_{1\cdot2}$ and $t_{2\cdot2}$ have the same sign. $x_{1\cdot2}$ and $x_{2\cdot2}$ may have different signs in case when

(5.10)
$$1 + \frac{v_{2\cdot1}}{v_{3\cdot2}} < 0$$

From the inequalities (5.9), (5.10), it follows that

(5.11)
$$-c < v_{2\cdot1} < -v_{3\cdot2}$$

The theorem follows from the equation (5.11). □

The theorem 5.1 can be stated differently.

COROLLARY 5.2. *Let the reference frame S_3 moves with speed $v_{3\cdot2}$*
$$0 < v_{3\cdot2} < c$$
relative to the reference frame S_2 in the direction of the increasing x. Let the reference frame S_1 moves with speed $v_{1\cdot2}$
$$0 < v_{1\cdot2} < c$$
relative to the reference frame S_2 in the direction of the increasing x. Since
$$v_{3\cdot2} < v_{1\cdot2}$$
then the reference frame S_3 moves relative to the reference frame S_1 in the direction of the decreasing x. □

REMARK 5.3. Let $|v_{3\cdot2}| \le c$. We assume[10]
$$v_{2\cdot1} = c(1 - a) \quad 2 \le a \le 0$$
$$v_{3\cdot2} = c(1 - b) \quad 2 \le b \le 0$$

Then the equation (5.5) has form

(5.12)
$$v_{3\cdot1} = \frac{c((1-b)+(1-a))}{1 + \frac{c^2(1-b)(1-a)}{c^2}} = c\frac{2-b-a}{2-b-a+ab}$$

Therefore $|v_{3\cdot1}| \le c$. □

---

[10]When assessing the value of speed $v_{3\cdot1}$, I follow to [1], §5.

THEOREM 5.4. *Let*

(5.13)
$$v_{3 \cdot 2} > c$$

(5.14)
$$-c < v_{2 \cdot 1} < c$$

Since the speed $|v_{2 \cdot 1}|$ *of the reference frame* S_2 *relative to the reference frame* S_1 *in the direction of the decreasing* x *satisfies the inequality*

(5.15)
$$\frac{c^2}{v_{3 \cdot 2}} < |v_{2 \cdot 1}| < c$$

then the reference frame S_3 *moves relative to the reference frame* S_1 *in the direction of the decreasing* x.

PROOF. From the equation (5.7) and the inequalities (5.13), (5.14), it follows that $x_{1 \cdot 2}$ and $x_{2 \cdot 2}$ have the same sign. $t_{1 \cdot 2}$ and $t_{2 \cdot 2}$ may have different signs in case when

(5.16)
$$1 + \frac{v_{2 \cdot 1} v_{3 \cdot 2}}{c^2} < 0$$

From the inequality (5.16), it follows that

(5.17)
$$-c < v_{2 \cdot 1} < -\frac{c^2}{v_{3 \cdot 2}}$$

The condition (5.15) follows from the condition (5.17). □

REMARK 5.5. The reference frame S_3 has a constant velocity in the direction of the increasing x of the reference frame S_1. With increasing x on the world line of the reference frame S_3 the coordinate t decreases. However the observer of the reference frame S_1 can sense values of the coordinate t only in ascending order. Therefore, the observer of the reference frame S_1 sees the movement of the reference frame S_3 in the direction of the decreasing x. □

The theorem 5.4 in view of the remark 5.5 can be stated differently.

COROLLARY 5.6. *Let the reference frame* S_3 *moves with speed* $v_{3 \cdot 2} > c$ *relative to the reference frame* S_2 *in the direction of the increasing* x. *Let the reference frame* S_1 *moves with speed* $v_{1 \cdot 2}$
$$\frac{c^2}{v_{3 \cdot 2}} < v_{1 \cdot 2} < c$$
relative to the reference frame S_2 *in the direction of the increasing* x. *Then the reference frame* S_3 *moves relative to the reference frame* S_1 *in the direction of the decreasing* x. □

REMARK 5.7. From the theorem 5.4 and the equation (5.5), it follows that provided
$$v_{2 \cdot 1} = -\frac{c^2}{v_{3 \cdot 2}}$$
speed $v_{3 \cdot 1}$ becomes infinitely large. This corresponds to the case when the world line of the reference frame S_3 coincides with the axis x in the reference frame S_1.

From the remark 5.3, it follows that $|v_{3 \cdot 1}| > c$. □

6. Tracking of Movement

To better understand the meaning of the remark 5.7, we consider how the observer of the reference frame S_2 tracks the movement of the reference frame S_3.

We start from consideration of procedure similar to Schild's ladder ([6], box 10.2, pp. 248, 249, box 16.4, p. 397). In the case

$$0 < v_{3.2} < c$$

this procedure takes the following form.

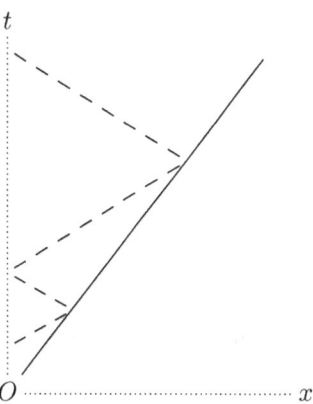

However, since

$$v_{3.2} > c$$

then the observer of the reference frame S_2 cannot use this procedure, since twice reflected ray does not meet the observer of the reference frame S_3.

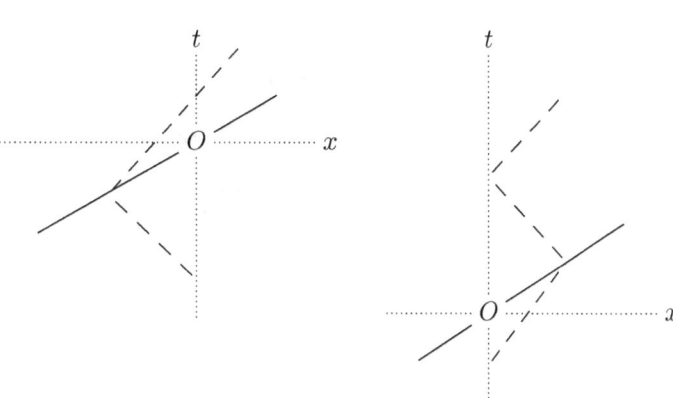

We can modify the considered procedure. For instance, the observer of the reference frame S_2 periodically sends a signal without waiting for an answer to the previous signal. By analogy with the known method of measuring distances we call this procedure echolocation.

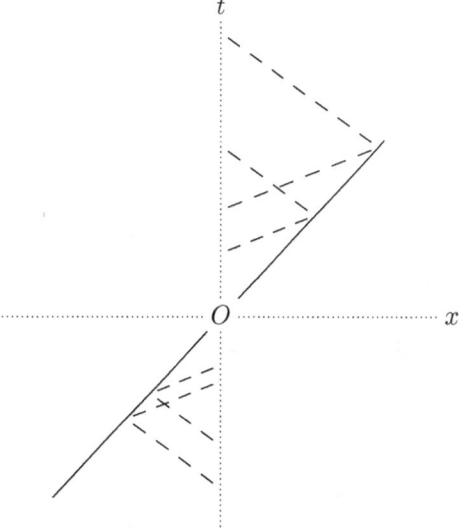

We consider the echolocation in the case

$$0 < v_{3.2} < c$$

At any given time, the observer of the reference frame S_2 can send ray along the axis of x in any direction. Since the angle between the world line of the reference frame S_3 and axis of t is less than the angle between the world line of the ray and axis of t, then only one ray crosses the world line of the reference frame S_3 in the future.

THEOREM 6.1. *Let the observer of the reference frame S_3 approaches the observer of the reference frame S_2 with speed*

$$0 < v_{3\cdot2} < c$$

Let

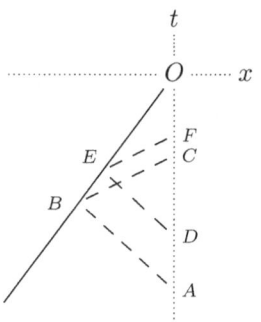

(6.1) $$\Delta_1 t_2 = t_{2\cdot D} - t_{2\cdot A}$$

be the time interval between rays emitted by the observer of the reference frame S_2. Let

(6.2) $$\Delta_2 t_2 = t_{2\cdot F} - t_{2\cdot C}$$

be the time interval in the reference frame S_2 between reception of reflected rays. Then

(6.3) $$\Delta_2 t_2 = \frac{c - v_{3\cdot2}}{c + v_{3\cdot2}} \Delta_1 t_2$$

PROOF. Since $AB \parallel DE$, then triangles ABO, DEO are similar. Therefore,

(6.4) $$\frac{t_{2\cdot D}}{t_{2\cdot A}} = \frac{DO}{AO} = \frac{EO}{BO} = \frac{t_{2\cdot E}}{t_{2\cdot B}}$$

Since $BC \parallel EF$, then triangles BCO, EFO are similar. Therefore,

(6.5) $$\frac{t_{2\cdot F}}{t_{2\cdot C}} = \frac{FO}{CO} = \frac{EO}{BO} = \frac{t_{2\cdot E}}{t_{2\cdot B}}$$

From equations (6.4), (6.5), it follows that

(6.6) $$\frac{t_{2\cdot D}}{t_{2\cdot A}} = \frac{t_{2\cdot F}}{t_{2\cdot C}}$$

From equations (6.1), (6.2), it follows that

(6.7) $$t_{2\cdot D} = t_{2\cdot A} - \Delta_1 t_2$$
$$t_{2\cdot F} = t_{2\cdot C} - \Delta_2 t_2$$

From equations (6.6), (6.7), it follows that

(6.8) $$\frac{t_{2\cdot A} - \Delta_1 t_2}{t_{2\cdot A}} = \frac{t_{2\cdot C} - \Delta_2 t_2}{t_{2\cdot C}}$$

$$1 - \frac{\Delta_1 t_2}{t_{2\cdot A}} = 1 - \frac{\Delta_2 t_2}{t_{2\cdot C}}$$

From the equation (6.8), it follows that

(6.9) $$\Delta_2 t_2 = \frac{t_{2\cdot C}}{t_{2\cdot A}} \Delta_1 t_2$$

Since the observer of the reference frame S_3 approaches the observer of the reference frame S_2 with speed

$$0 < v_{3\cdot2} < c$$

then

(6.10) $$v_{3\cdot2} t_{2\cdot B} = -c(t_{2\cdot B} - t_{2\cdot A})$$

From the equation (6.10), it follows that

(6.11) $$t_{2\cdot B} = \frac{c}{v_{3\cdot2} + c} t_{2\cdot A}$$

Since

(6.12) $$t_{2 \cdot C} - t_{2 \cdot B} = t_{2 \cdot B} - t_{2 \cdot A}$$

then from equations (6.11), (6.12) it follows that

(6.13) $$t_{2 \cdot C} = 2t_{2 \cdot B} - t_{2 \cdot A} = 2\frac{c}{c - v_{3 \cdot 2}} t_{2 \cdot A} - t_{2 \cdot A} = \frac{c + v_{3 \cdot 2}}{c - v_{3 \cdot 2}} t_{2 \cdot A}$$

The equation (6.3) follows from equations (6.9), (6.13). □

COROLLARY 6.2. *Let the observer of the reference frame S_3 approaches the observer of the reference frame S_2 with speed*

$$0 < v_{3 \cdot 2} < c$$

Then the time interval in the reference frame S_2 between reception of reflected rays is less than the time interval between rays emitted by the observer of the reference frame S_2. □

THEOREM 6.3. *Let the observer of the reference frame S_3 moves away from the observer of the reference frame S_2 with speed*

$$0 < v_{3 \cdot 2} < c$$

Let

(6.14) $$\Delta_1 t_2 = t_{2 \cdot D} - t_{2 \cdot A}$$

be the time interval between rays emitted by the observer of the reference frame S_2. Let

(6.15) $$\Delta_2 t_2 = t_{2 \cdot F} - t_{2 \cdot C}$$

be the time interval in the reference frame S_2 between reception of reflected rays. Then

(6.16) $$\Delta_2 t_2 = \frac{c + v_{3 \cdot 2}}{c - v_{3 \cdot 2}} \Delta_1 t_2$$

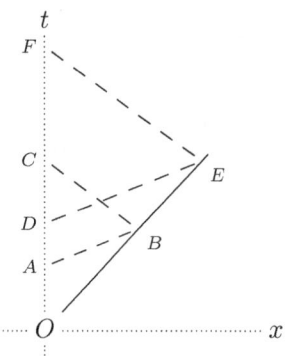

PROOF. Since $AB \parallel DE$, then triangles ABO, DEO are similar. Therefore,

(6.17) $$\frac{t_{2 \cdot D}}{t_{2 \cdot A}} = \frac{DO}{AO} = \frac{EO}{BO} = \frac{t_{2 \cdot E}}{t_{2 \cdot B}}$$

Since $BC \parallel EF$, then triangles BCO, EFO are similar. Therefore,

(6.18) $$\frac{t_{2 \cdot F}}{t_{2 \cdot C}} = \frac{FO}{CO} = \frac{EO}{BO} = \frac{t_{2 \cdot E}}{t_{2 \cdot B}}$$

From equations (6.17), (6.18), it follows that

(6.19) $$\frac{t_{2 \cdot D}}{t_{2 \cdot A}} = \frac{t_{2 \cdot F}}{t_{2 \cdot C}}$$

From equations (6.14), (6.15), it follows that

(6.20) $$t_{2 \cdot D} = t_{2 \cdot A} - \Delta_1 t_2$$
$$t_{2 \cdot F} = t_{2 \cdot C} - \Delta_2 t_2$$

From equations (6.19), (6.20), it follows that

$$\frac{t_{2 \cdot A} - \Delta_1 t_2}{t_{2 \cdot A}} = \frac{t_{2 \cdot C} - \Delta_2 t_2}{t_{2 \cdot C}}$$

(6.21)

$$1 - \frac{\Delta_1 t_2}{t_{2 \cdot A}} = 1 - \frac{\Delta_2 t_2}{t_{2 \cdot C}}$$

From the equation (6.21), it follows that

(6.22)
$$\Delta_2 t_2 = \frac{t_{2 \cdot C}}{t_{2 \cdot A}} \Delta_1 t_2$$

Since the observer of the reference frame S_3 moves away from the observer of the reference frame S_2 with speed

$$0 < v_{3 \cdot 2} < c$$

then

(6.23)
$$v_{3 \cdot 2} t_{2 \cdot B} = c(t_{2 \cdot B} - t_{2 \cdot A})$$

From the equation (6.23), it follows that

(6.24)
$$t_{2 \cdot B} = \frac{c}{c - v_{3 \cdot 2}} t_{2 \cdot A}$$

Since

(6.25)
$$t_{2 \cdot C} - t_{2 \cdot B} = t_{2 \cdot B} - t_{2 \cdot A}$$

then from equations (6.24), (6.25) it follows that

(6.26)
$$t_{2 \cdot C} = 2 t_{2 \cdot B} - t_{2 \cdot A} = 2 \frac{c}{v_{3 \cdot 2} + c} t_{2 \cdot A} - t_{2 \cdot A} = \frac{c - v_{3 \cdot 2}}{c + v_{3 \cdot 2}} t_{2 \cdot A}$$

The equation (6.16) follows from equations (6.22), (6.26). $\qquad \square$

COROLLARY 6.4. *Let the observer of the reference frame S_3 moves away from the observer of the reference frame S_2 with speed*

$$0 < v_{3 \cdot 2} < c$$

Then the time interval in the reference frame S_2 between reception of reflected rays is longer than the time interval between rays emitted by the observer of the reference frame S_2. $\qquad \square$

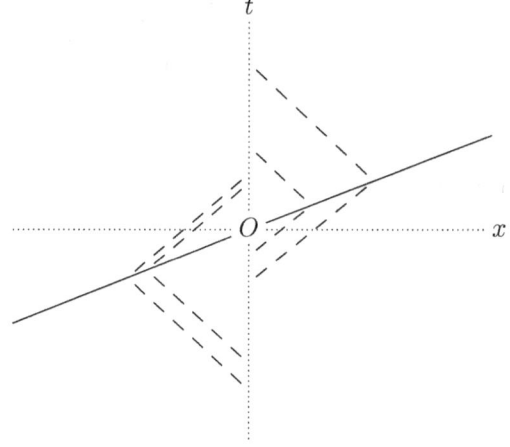

If

$$v_{3 \cdot 2} > c$$

then the angle between the world line of the reference frame S_3 and axis of t is larger than the angle between the world line of the ray and axis of t. So both rays emitted by the observer of the reference frame S_2 at any time prior to the meeting with the observer of the reference frame S_3 will cross the world line of the reference frame S_3. The observer of the reference frame S_2 receives reflected rays after the meeting with the observer of the reference frame S_3.

THEOREM 6.5. *Let the observer of the reference frame S_3 approaches the observer of the reference frame S_2 with speed*

$$v_{3\cdot2} > c$$

Let

$$\Delta_1 t_2 = t_{2\cdot D} - t_{2\cdot A}$$

be the time interval between rays emitted by the observer of the reference frame S_2. Let

$$\Delta_2 t_2 = t_{2\cdot F} - t_{2\cdot C}$$

be the time interval in the reference frame S_2 between reception of reflected rays. Then

$$\Delta_2 t_2 = \frac{c - v_{3\cdot2}}{c + v_{3\cdot2}} \Delta_1 t_2$$

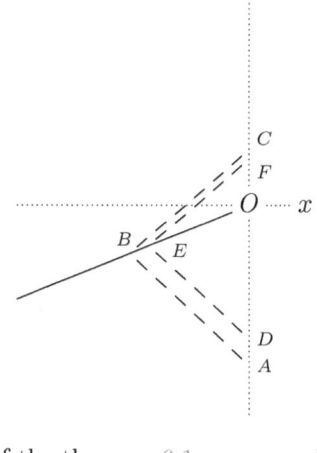

PROOF. The proof of the theorem is the same as the proof of the theorem 6.1. □

COROLLARY 6.6. *Let the observer of the reference frame S_3 approaches the observer of the reference frame S_2 with speed*

$$v_{3\cdot2} > c$$

Then the time interval in the reference frame S_2 between reception of reflected rays is less than the time interval between rays emitted by the observer of the reference frame S_2. □

THEOREM 6.7. *Let the observer of the reference frame S_3 moves away from the observer of the reference frame S_2 with speed*

$$v_{3\cdot2} > c$$

Let

$$\Delta_1 t_2 = t_{2\cdot D} - t_{2\cdot A}$$

be the time interval between rays emitted by the observer of the reference frame S_2. Let

$$\Delta_2 t_2 = t_{2\cdot F} - t_{2\cdot C}$$

be the time interval in the reference frame S_2 between reception of reflected rays. Then

$$\Delta_2 t_2 = \frac{c + v_{3\cdot2}}{c - v_{3\cdot2}} \Delta_1 t_2$$

PROOF. The proof of the theorem is the same as the proof of the theorem 6.3. □

COROLLARY 6.8. *Let the observer of the reference frame S_3 moves away from the observer of the reference frame S_2 with speed*

$$v_{3\cdot2} > c$$

Then the time interval in the reference frame S_2 between reception of reflected rays is longer than the time interval between rays emitted by the observer of the reference frame S_2. □

The observer of the reference frame S_2 receives reflected rays in the order opposite to the order of emission. So the observer of the reference frame S_2 may observe the reference frame S_3 as two reference frames moving in different directions. However, since the observer of the reference frame S_2 modulates radiated signals by time of radiation, then he can uniquely reconstruct the motion of the reference frame S_3.

According to statements 6.2, 6.6, regardless of the value of speed $v_{3\cdot2}$, if the observer of the reference frame S_3 approaches the observer of the reference frame S_2, then interval between reflected signals is less then interval between radiated signals. According to statements 6.4, 6.8, if the the observer of the reference frame S_3 moves away from the observer of the reference frame S_2, then interval between reflected signals is larger than interval between radiated signals. This phenomenon is similar to Doppler effect ([3], pages 34-7, 34-8, [4], pages 145 - 147). So this phenomenon is also called Doppler effect.

Consider reference frames S_1, S_2 such that $v_{3.1} < 0$, $v_{3.2} > 0$. Let A_1, A_2 be events on the world line of the reference frame S_3.

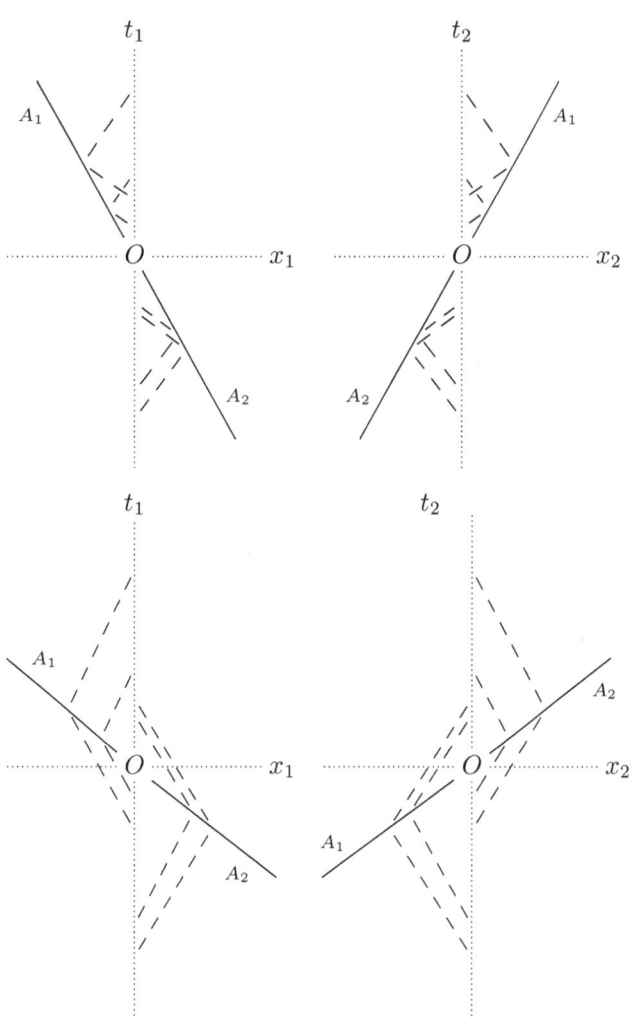

Let $v_{3.2} < c$. It is easy to see that the reference frame S_3 moves in direction of the decreasing x of the reference frame S_1 and in direction of the increasing x of the reference frame S_2. From the proof of the theorem 5.1, it follows that the order of events A_1, A_2 along the axis of x changes; however, order of these events along the axis of t is preserved. This indicates the preservation of causality.

Let $v_{3.2} > c$. According to the observed Doppler effect, the reference frame S_3 moves in direction of the decreasing x of the reference frame S_1 and in direction of the increasing x of the reference frame S_2. From the proof of the theorem 5.4, it follows that the order of events A_1, A_2 along the axis of t changes; however, order of these events along the axis of x is preserved.

Therefore, the observer of the reference frame S_1 observes movement of the reference frame S_3 from the event A_2 to the event A_1 and the observer of the reference frame S_2 observes movement of the reference frame S_3 from the event A_1 to the event A_2. This indicates a violation of causality for the reference frame S_3.

Echolocation can be used not only to track motion with constant velocity. However, this method is effective only at short distances. So we can modify the method of echolocation. The observer of the reference frame S_3 sends signals to the observer of the reference frame S_1 at specified intervals. Radiation can be continuous; but then the frequency of radiation should be specified.[11]

---

[11]Such communication NASA implemented to determine the trajectory of Pioneer 10 space probe ([7, 8]).

Since all triangles formed by the axis of t, the world line of the reference frame S_3 and ray are similar, the diagrams in the event space are similar to the diagram of echolocation.

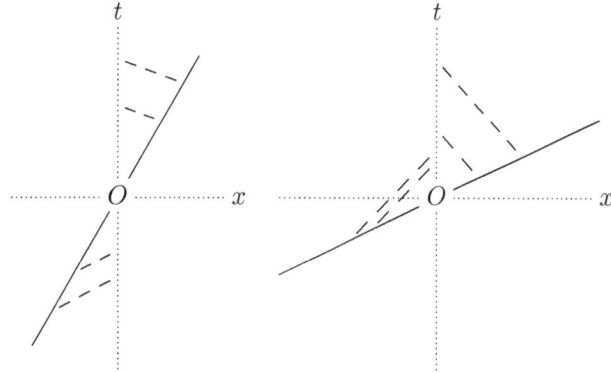

7. References

[1] Albert Einstein, On the Electrodynamics of Moving Bodies, 1905,
The Principle of Relativity: A Collection of Original Memoirs on the Special and General Theory of Relativity , 37 - 65,
Courier Dover Publications, 1952; ISBN-13: 978-0486600819
Zur Elektrodynamik der bewegter Körper. Ann. Phys., 1905, 17, 891-921.

[2] Albert Einstein, On the Relativity Principle and the Conclusions Drawn from It, 1907,
The Collected Papers of Albert Einstein, Volume 2: The Swiss Years: Writings, 1900-1909. English translation. 252 - 311.
Anna Beck, translator; Peter Havas, consultant. Princeton University Press, 1989; ISBN-13: 9780691085494
Über das Relativitätsprinzip und die aus demselben gezogenen Folgerungen. Jahrb. d. Radioaktivität u. Elektronik, 1907, 4, 411-462.

[3] Richard Phillips Feynman, Robert B. Leighton, Matthew Linzee Sands. The Feynman lectures on physics: Volume 1. Mainly Mechanics, Radiation, and Heat. Addison-Wesley, 1965.

[4] James Shipman, Jerry D. Wilson and Aaron Todd. Introduction to Physical Science. Cengage Learning, 2009; ISBN 0538731877.

[5] Walter Greiner, Joachim Reinhardt. Quantum Electrodynamics. Springer, 2009.

[6] Charles W. Misner, Kip S. Thorne, John Archibald Wheeler. Gravitation. W. H. Freeman and Company, San Francisco, 1973.

[7] J. D. Anderson, P. A. Laing, E. L. Lau, A. S. Liu, M. M. Nieto, and S. G. Turyshev, Indication, from Pioneer 10/11, Galileo, and Ulysses Data, of an Apparent Anomalous, Weak, Long-Range Acceleration, Phys. Rev. Lett. 81, 2858, (1998), eprint arXiv:gr-qc/9808081 (1998)

[8] J. D. Anderson, P. A. Laing, E. L. Lau, A. S. Liu, M. M. Nieto, and S. G. Turyshev, Study of the anomalous acceleration of Pioneer 10 and 11, Phys. Rev. D 65, 082004, 50 pp., (2002), eprint arXiv:gr-qc/0104064 (2001)

[9] The OPERA Collaboration. Measurement of the neutrino velocity with the OPERA detector in the CNGS beam. eprint arXiv:1109.4897 [hep-ex] (2011)

[10] Aleks Kleyn, Reference Frame in General Relativity, eprint arXiv:gr-qc/0405027 (2008)